Bitwarden Essentials

A Practical Guide to Password Security

L. Phillips

PRAIRIE LIGHT PRESS

ESTD. 2025

Helping readers move from the unknown to the known

Published by Prairie Light Press

Distributed by IngramSpark Press

Second Edition

English Edition

Publication Identifiers

ISBN (Hardcover) 979-8-9969802-1-5
ISBN (Paperback) 979-8-9978724-0-3
ASIN (Kindle) B0G765KDBT

Disclaimer

The information contained in this book is provided for educational purposes only.

The author makes no representations or warranties regarding the accuracy, completeness, or suitability of the information contained herein. The techniques and procedures described may not be appropriate for all systems or environments.

The author shall not be held liable for any loss, damage, or disruption caused by reliance on the information provided in this book.

Readers are responsible for evaluating and applying the information in accordance with their own circumstances and applicable laws.

No Affiliation Statement

Any product names, logos, brands, or trademarks referenced in this book are the property of their respective owners. Their use does not imply endorsement.

Table of Contents

About This Book 1

Introduction: Take Control of Your Passwords 3

Chapter 1: Why Your Passwords Need Help 9

Chapter 2: Getting Started with Bitwarden 17

Chapter 3: Installing Bitwarden Everywhere 27

Chapter 4: Adding Your Passwords 39

Chapter 5: Using Bitwarden Every Day 51

Chapter 6: Keeping Things Organized 62

Chapter 7: Sharing Passwords with Family 72

Chapter 8: Staying Secure 84

Chapter 9: When Things Go Wrong (Quick Fixes)........ 96

Book Preview – Guard Your Money 118

If you want to get started immediately, see the Quick-Start Guide at the end of the book

About This Book

Bitwarden Essentials is your friendly guide to taking control of your passwords. Whether you're tired of clicking "Forgot Password" or you're using the same password everywhere (we've all done it), this book will help you get secure without the complexity.

What You'll Learn:

- Set up Bitwarden in under 30 minutes
- Store all your passwords securely in one place
- Log into websites with just one click
- Share passwords safely with family
- Stay protected from hackers and breaches
- Troubleshoot common problems

Who This Book Is For:

- Regular people who aren't tech experts
- Anyone with more than 10 online accounts
- Families who share streaming services and WiFi passwords
- People who want better security without complexity

What You Don't Need:

- Technical knowledge
- Computer science degree
- Hours of time
- Expensive software

Just 2-3 hours, a willingness to learn, and you'll have password security figured out for life.

Let's get started!

Introduction: Take Control of Your Passwords

You Need This Book If…

- You use the same password for multiple websites (we all do it, but it's risky)
- You've clicked "Forgot Password" more times than you'd like to admit
- You have a notebook full of passwords or a sticky note on your monitor
- You're tired of resetting passwords every time you need to log in
- You've heard about password managers but don't know where to start
- You want better security without making your life more complicated

If any of these sound familiar, you're in the right place.

What This Book Will Do for You

By the end of this book, you'll:

- Have all your passwords stored securely in one place

- Never forget another password

- Create strong passwords without thinking about it

- Log into websites with just one click

- Share Netflix passwords with family safely

- Know exactly what to do if a website gets hacked

- Sleep better knowing your accounts are protected

And here's the best part: **it's easier than what you're doing now.**

What Is Bitwarden?

Think of Bitwarden as a super-secure vault that stores all your passwords. Instead of remembering 100 different passwords, you only remember one master password that unlocks your vault. Bitwarden does the rest.

It's like this:

- Old way: Remember 100 passwords (impossible)

→ Use same password everywhere (dangerous)

→ Get hacked - New way: Remember 1 password

→ Bitwarden remembers the other 99

→ You're secure

Where it works:

- On your computer (Windows, Mac, Linux)

- In your web browser (Chrome, Firefox, Safari, Edge)

- On your phone (iPhone, Android)

- On your tablet - Everywhere you need it

Best of all: It's free for most people. The basic version does everything you need.

Why Bitwarden?

There are other password managers out there, so why Bitwarden?

It's trustworthy:

- Open source (security experts can verify it's safe)

- Used by millions of people

- Regularly audited by security professionals

- Your passwords are encrypted so even Bitwarden can't see them

It's affordable:

- Free version is genuinely useful (not a limited trial)

- Premium version costs $10 per *year* (less than a dollar a month)

- Family plan for 6 people: $40 per year

It works everywhere: - All your devices stay in sync - Available on every platform - No artificial limits on how many devices you can use

It's yours:

- You own your data

- You can export everything anytime

- No lock-in, no tricks

What Makes This Book Different

This isn't a technical manual. This is a practical guide written for regular people.

You won't find:

- Complicated technical jargon

- Advanced features you'll never use

- Scary security lectures

- Assumptions that you're a computer expert

You will find:

- Simple, step-by-step instructions

- Real-world examples

- Screenshots and visual guides

- Tips that actually help

- Answers to questions you didn't know you had

How to Use This Book

If you're brand new: Read it straight through, chapters 1-9. Do the exercises. Set up Bitwarden as you go. By the end, you'll be a confident password manager user.

If you already use Bitwarden: Skip to the chapters that interest you. Each chapter works independently.

If you just want to get started fast: Read Chapters 2-4, then come back to the rest later.

If something goes wrong: Jump to Chapter 9 (Common Problems) for quick fixes.

What You'll Need

- A computer, phone, or tablet with internet access
- An email address
- About 2-3 hours to set everything up
- Willingness to try something new

That's it. No special skills required.

A Word About Security

You might be thinking: "Isn't putting all my passwords in one place risky?"

Fair question. Here's the truth:

What you're doing now is riskier:

- Using the same password everywhere means one breach exposes everything

- Weak passwords are easy to hack

- Written passwords can be found by anyone

- Browser-saved passwords aren't encrypted properly

Bitwarden is designed for this:

- Military-grade encryption protects your passwords

- Even if someone hacks Bitwarden's servers, they get useless encrypted data

- Only you can unlock your passwords

- It's far more secure than sticky notes or memory

Think of it this way: Would you rather keep cash scattered around your house, or in a safe? Bitwarden is the safe.

Let's Get Started

Password security doesn't have to be complicated, expensive, or time-consuming. With Bitwarden, it's actually easier than what you're doing now.

Ready to take control of your passwords?

Let's do this.

Note: This book covers everything you need to use Bitwarden confidently. We skip the ultra-advanced features and focus on what matters: getting you secure and making your digital life easier.

If you later want to explore advanced topics like self-hosting your own password server, check out "Bitwarden: The Complete Reference" for the full technical deep dive. But for most people, this Essentials guide is all you'll ever need.

Now, turn the page and let's start securing your digital life.

Chapter 1: Why Your Passwords Need Help

The Problem We All Have

Here's what probably happens to you:

Scenario 1: You need to create a new account. The website says "password must be 8 characters with a number and special character." You think of your usual password, add a "1" at the end, and hope you remember it later.

Scenario 2: You try to log into a website you haven't used in months. Wrong password. Try another. Wrong again. Click "Forgot Password." Wait for email. Reset password. Forget it again next time.

Scenario 3: You hear that a website got hacked. You used the same password there that you use on 10 other sites. Now you have to change all 10. But you probably won't.

Sound familiar? You're not alone, and you're not doing anything wrong. The system is broken.

Why Passwords Are Impossible

The average person has 100+ online accounts.

Let's be honest: Nobody can remember 100 different complex passwords. It's humanly impossible. So we do what makes sense:

1. **Use the same password everywhere**
   - Easy to remember

- Seems practical
- But super dangerous

2. **Use simple passwords**
   - "password123"
   - Your pet's name
   - Your birthday
   - All easily guessed
3. **Write them down**
   - Sticky notes
   - Notebooks
   - Phone notes
   - Anyone can find them
4. **Use patterns**
   - Facebook2024
   - Gmail2024
   - Amazon2024
   - Easy to crack once someone knows the pattern

None of these are your fault. The problem is that websites expect you to have a superhuman memory.

What Happens When Passwords Fail

When you reuse passwords:

Imagine you use the same password for Netflix, your email, and your bank. A small gaming forum you signed up for years ago gets hacked. Hackers now have your email and password.

They try that combination on:

- Gmail ✓ They're in - Bank website ✓ They're in

- Amazon ✓ They're in

- Everything using that password ✓ They're in

One website's mistake just exposed your entire digital life.

Real-world impact:

- Stolen money from bank accounts

- Unauthorized purchases

- Locked out of your own accounts

- Identity theft

- Spam sent to all your contacts

- Private photos exposed

This happens to millions of people every year. Not because they're careless, but because the current system is impossible to use securely.

What Makes a Good Password?

A strong password is:

- - **Long** - At least 12 characters (longer is better)
- - **Random** - Not a word from dictionary
- - **Unique** - Different for every website
- - **Complex** - Mix of letters, numbers, symbols

Example of a weak password:

```
fluffy2024
```

- Common word
- Predictable number
- Easy to guess
- Takes seconds to crack

Example of a strong password:

```
X7$mK9#pL2@nQ5*j
```

- Completely random
- Mix of everything
- Would take billions of years to crack
- Impossible to remember (that's okay!)

The Catch-22

To be secure, you need:

- A different password for every website

- Each password should be long and random

- You need to remember all of them

This is impossible.

Which is why password managers exist.

Enter: Password Managers

A password manager solves the impossible problem:

What it does:

1. You remember ONE strong password (the master password)

2. The password manager remembers all your other passwords

3. When you visit a website, it fills in the password for you

4. Everything is encrypted and secure

It's like:

- Having a super-secure vault for all your passwords

- The vault has one key (your master password)

- Only you have the key

- Inside the vault are all your other passwords

The magic:

- You don't have to remember complex passwords

- You don't have to type them

- Each website gets a different password

- If one website gets hacked, others are still safe

How Bitwarden Keeps You Safe

Military-grade encryption: Your passwords are scrambled using the same encryption that protects government secrets. Even if someone steals the encrypted data, it's useless without your master password.

Zero-knowledge: Even Bitwarden's employees can't see your passwords. The encryption happens on your device before anything goes to their servers.

You're in control:

- Only you know your master password

- Only you can unlock your vault

- You can export everything anytime

- You own your data

It just works:

- Syncs across all your devices

- Fills in passwords automatically

- Generates strong passwords for you

- Tells you if a password has been compromised

What You'll Do Differently

Before Bitwarden:

1. Try to log into website

2. Wrong password

3. Click "Forgot Password"

4. Reset password

5. Hope you remember it next time

6. (Spoiler: You won't)

With Bitwarden:

1. Visit website
2. Bitwarden fills in your password
3. You're logged in
4. Done

Creating new accounts:

Before:

- Think of password
- Add number at end
- Hope it's strong enough
- Hope you remember it

With Bitwarden:

- Click "generate password"
- Bitwarden creates: `K9$mX7#pL2@Q5*jn`
- Bitwarden saves it automatically
- You never even see it

Common Concerns (Answered)

"What if I forget my master password?" You write it down and keep it somewhere safe (like a real safe). We'll cover this in detail.

"What if Bitwarden gets hacked?" Your passwords are encrypted. Hackers would get useless scrambled data they can't decrypt.

"Isn't putting all passwords in one place risky?" Less risky than using the same password everywhere, which is what most people do now.

"Sounds complicated." It's actually easier than what you're doing now. Bitwarden does the hard work for you.

"Costs money?" The free version does everything most people need. Premium is $10 per year if you want extra features.

What's Next

In the next chapter, we'll create your Bitwarden account and set up your master password. This is the most important password you'll ever create, so we'll take our time and get it right.

By the end of Chapter 2, you'll have Bitwarden set up and ready to use.

By the end of this book, you'll wonder how you ever managed passwords any other way.

Let's get started.

Chapter 2: Getting Started with Bitwarden

What We'll Do in This Chapter

1. Create your Bitwarden account (5 minutes)
2. Create a strong master password you can actually remember (10 minutes)
3. Set up two-factor authentication for extra security (5 minutes)

Total time: About 20 minutes

Grab a cup of coffee and let's do this.

Creating Your Bitwarden Account

Figure 1 – Create Account – AI Generated

Step 1: Go to the Website

Open your web browser and go to: **bitwarden.com**

Click the big "Get Started" button.

Step 2: Enter Your Email

Use an email address you check regularly. This will be your username for logging into Bitwarden.

Important: Use a personal email (Gmail, Outlook, etc.), not a work email. You'll want access to this even if you change jobs.

Step 3: Enter Your Name

Just your first and last name. This is just for personalizing the interface—it's not critical.

Step 4: Create Your Master Password

STOP. This is the most important password you'll ever create. Don't rush it.

We're going to spend the next section making sure you get this right.

Your Master Password: The Key to Everything

Your master password is special because:

- It's the ONLY password you'll need to remember

- It unlocks ALL your other passwords

- Bitwarden can't reset it if you forget it

- You'll type it every day, so it needs to be memorable

- But it also needs to be strong

Figure 2 – Set Strong Password – AI Generated

The Challenge:

It needs to be both strong AND memorable.

The Solution: We'll use a method that makes this easy.

The Simple Passphrase Method

Instead of a complicated password like `Tr!cky$P@ssw0rd`, we'll use random words:

```
correct-horse-battery-staple
```

Why this works:

- Four random words = very strong

- Easy to remember (create a mental picture)

- Easy to type

- Harder to crack than complex short passwords

How to create yours:

Step 1: Generate Random Words

Go to: **useapassphrase.com** or use these lists:

Option A - Pick 4 random words from this list:

```
ocean, piano, rocket, garden, coffee, window,
dragon, sunset, camera, puzzle, mountain, kitchen,
thunder, bicycle, compass, library, marble, forest
```

Option B - Use a random word generator website

Step 2: Create Your Phrase

Let's say you picked: `dragon, coffee, mountain, piano`

Your master password: `dragon-coffee-mountain-piano`

Step 3: Make It Memorable

Create a vivid mental image: "A dragon drinking coffee on a mountain while playing piano"

The weirder the image, the more memorable it is.

Step 4: Practice Typing It

Type your new master password 10 times right now:

1. `dragon-coffee-mountain-piano`

2. `dragon-coffee-mountain-piano`

3. (keep going...)

Get it into muscle memory.

Step 5: Write It Down

Yes, really. Write it on paper and put it somewhere safe:

- In a locked drawer

- In a home safe

- With important documents

This is your backup. If you forget it, you're locked out forever.

Back to Creating Your Account

Now that you have your master password:

Step 4 (continued): Enter Your Master Password

Type your new passphrase in the "Master Password" field.

Step 5: Confirm Your Master Password

Type it again in the "Re-type Master Password" field.

Be careful—if you mistype it, you'll create an account with a password you don't actually know!

Step 6: Password Hint (Optional)

If you want, add a hint like: "Four words: mythical creature, morning drink, tall landform, musical instrument"

Don't write the actual words. The hint should jog YOUR memory without revealing the password to others.

I usually skip this field entirely.

Step 7: Create Account

Click "Submit" and you'll get a verification email.

Step 8: Verify Your Email

Check your email and click the verification link.

Congratulations! You now have a Bitwarden account.

Adding Two-Factor Authentication (2FA)

This adds an extra layer of security. Even if someone steals your master password, they can't get in without this second step.

What is 2FA?

When you log in, you need two things:

1. Your master password (something you know)

2. A code from your phone (something you have)

Setting it up (5 minutes):

Step 1: Download an Authenticator App

On your smartphone, install one of these FREE apps:

- **Authy** (recommended for beginners—has cloud backup)

- **Google Authenticator**

- **Microsoft Authenticator**

Step 2: Enable 2FA in Bitwarden

1. Log into vault.bitwarden.com
2. Click your email address (top right)
3. Click "Account Settings"
4. Click "Security"
5. Click "Two-step Login"
6. Click "Manage" next to "Authenticator App"

Step 3: Scan the QR Code

1. Open your authenticator app on your phone
2. Click "Add" or "+"
3. Point your camera at the QR code on screen
4. It adds Bitwarden to your authenticator app

Step 4: Enter the 6-Digit Code

Your authenticator app shows a 6-digit number. Type it into Bitwarden where it says "Verification Code"

Step 5: Save Your Recovery Code

Bitwarden shows you a recovery code. **This is critical.**

Write it down on paper and store it with your master password.

If you lose your phone, this code lets you get back in.

Without this code, if you lose your phone, you're locked out forever.

Click "Close" after you've saved the code.

Step 6: Test It

1. Log out of Bitwarden
2. Log back in
3. Enter your master password
4. When prompted, enter the 6-digit code from your authenticator app
5. Success!

What You've Accomplished

☑ Created a Bitwarden account

☑ Made a strong master password you can remember

☑ Wrote down your master password in a safe place

☑ Set up two-factor authentication

☑ Saved your recovery code

☑ Tested that everything works

Your account is now secure and ready to use.

Quick Reminders

Your Master Password:

- Only you know it

- Bitwarden can't reset it

- You have it written down somewhere safe

- You'll use it every day until it's muscle memory

Your 2FA:

- Adds extra security

- Recovery code is safely stored

- You tested it and it works

What's Next:

In Chapter 3, we'll install Bitwarden on your computer and phone so you can access your passwords everywhere.

Then in Chapter 4, we'll start adding your actual passwords to Bitwarden.

Take a break if you need one. You've done the hard part. Everything else is easier.

Common Questions

Q: What if I forget my master password?

A: Check where you wrote it down. If you really lose it, you'll need to create a new account and start over. There's no password reset (by design, for security).

Q: What if I lose my phone with the authenticator app?

A: Use your recovery code to get back in. Then set up 2FA again on your new phone.

Q: Can I change my master password later?

A: Yes, but only if you know the current one. Account Settings → Change Master Password.

Q: Do I need premium?

A: Not yet. The free version does everything you need to get started. We'll talk about premium features later.

Q: Is my data backed up?

A: Bitwarden stores your encrypted vault on their servers. We'll also show you how to export your own backup later.

Ready for Chapter 3? Let's install Bitwarden on your devices.

Chapter 3: Installing Bitwarden Everywhere

Where You'll Use Bitwarden

You'll install Bitwarden in two places:

1. **Your web browser** (for using passwords on websites)

2. **Your phone** (for using passwords in apps and mobile browsing)

That's it. Simple.

You can also install a desktop app if you want, but the browser extension does everything most people need.

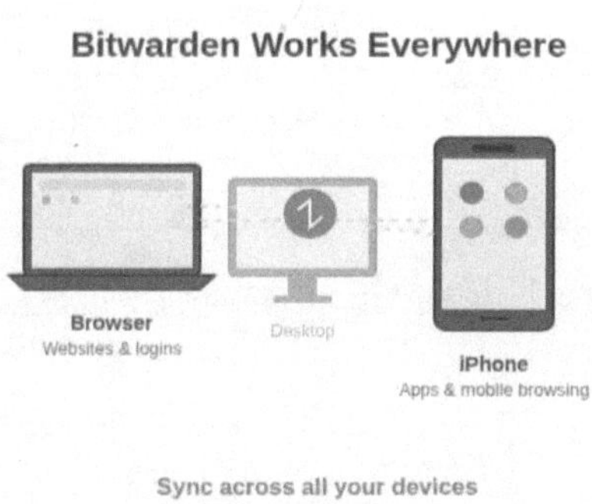

Figure 3 - Bitwarden Works Everywhere – AI Generated

Part 1: Installing in Your Browser

This is where you'll use Bitwarden most often—when browsing websites on your computer.

For Chrome Users

Step 1: Open Chrome

Step 2: Go to the Chrome Web Store

- Visit: **chrome.google.com/webstore**

- Or Google: "bitwarden chrome"

Step 3: Search for "Bitwarden"

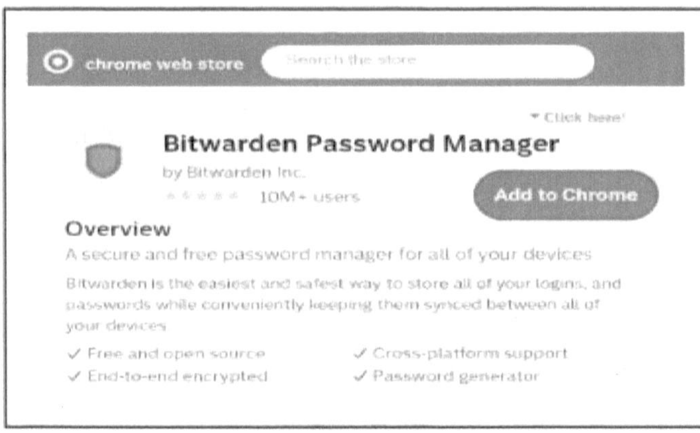

Figure 4 – Bitwarden Password Manager – Chrome – LP/PLP

Step 4: Click "Add to Chrome"

Step 5: Click "Add Extension"

Step 6: Look for the Bitwarden icon

You'll see a puzzle piece icon in the top right of Chrome. Click it, then click the pin icon next to Bitwarden to keep it visible.

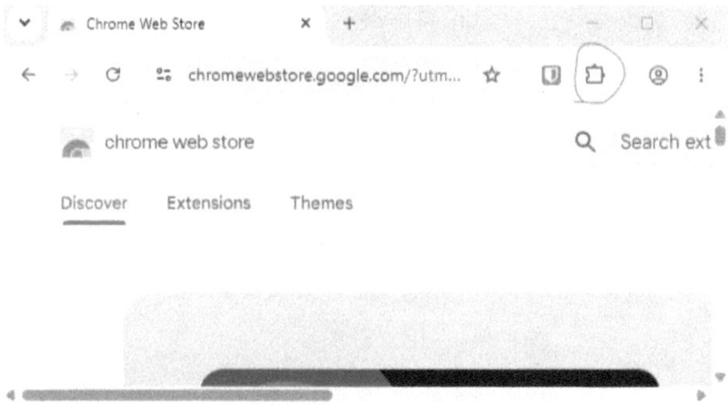

Figure 5 – Chrome & Bitwarden Extensions – LP/PLP

Step 7: Click the Bitwarden Icon

Step 8: Log In

- Enter your email

- Enter your master password

- Enter your 2FA code (from authenticator app)

Done! Bitwarden is now installed in Chrome.

For Firefox Users

Step 1: Open Firefox

Step 2: Go to Firefox Add-ons

- Visit: **addons.mozilla.org**

- Or click the menu (≡) → Add-ons and themes

Step 3: Search for “Bitwarden”

Step 4: Click “Add to Firefox”

Step 5: Click “Add” when prompted

Step 6: Click the Bitwarden icon in your toolbar

Step 7: Log in with your credentials

Done!

For Edge Users

Step 1: Open Microsoft Edge

Step 2: Go to Edge Add-ons - Visit:
microsoftedge.microsoft.com/addons

Step 3: Search for “Bitwarden”

Step 4: Click “Get”

Step 5: Click “Add Extension”

Step 6: Click the Bitwarden icon

Step 7: Log in

Done!

For Safari Users (Mac)

Safari is a bit different—the extension comes with the Mac desktop app.

Step 1: Download Bitwarden for Mac

- Visit: **bitwarden.com/download**

- Download “macOS Desktop”

Step 2: Install the app

- Open the downloaded file

- Drag Bitwarden to Applications

Step 3: Open Bitwarden from Applications

Step 4: Log in

Step 5: Enable Safari Extension

- Safari → Settings → Extensions

- Check the box next to Bitwarden

- Grant permissions

Done!

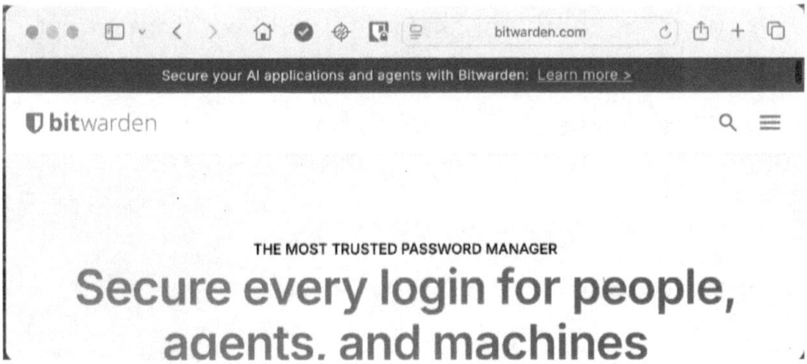

Figure 6 – Safari Extension Image – LP/PLP

Testing Your Browser Extension

Let’s make sure it works:

Test 1: Open the Extension

- Click the Bitwarden icon in your browser

- You should see "My Vault"

- It's currently empty (we'll fill it soon)

Test 2: Keyboard Shortcut

- Press `Ctrl + Shift + Y` (Windows/Linux)

- Or `Cmd + Shift + Y` (Mac)

- The Bitwarden popup should appear

Test 3: Try a Website

- Go to any website with a login (like Gmail)

- Click the Bitwarden icon

- You won't have any passwords saved yet (that's next chapter)

- But you should see the search box and "Add Item" button

If all three work, you're good to go!

Part 2: Installing on Your Phone

Now let's get Bitwarden on your phone for when you're mobile.

For iPhone/iPad

Step 1: Open the App Store

Step 2: Search for "Bitwarden"

Step 3: Look for "Bitwarden Password Manager" - Make sure it's by "Bitwarden.com"

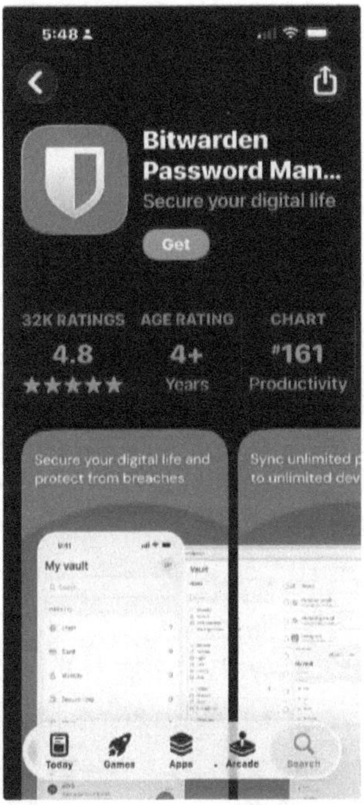

Figure 7 – Bitwarden App Store Image – LP/PLP

Step 4: Tap "Get" → Install

Step 5: Open the Bitwarden App

Step 6: Tap "Log In"

Step 7: Enter your credentials

- Email - Master password

- 2FA code

Step 8: Enable Auto-Fill This lets Bitwarden fill passwords in apps and Safari.

1. Open iPhone Settings (not Bitwarden settings)
2. Scroll down to “Passwords”
3. Tap “Password Options”
4. Enable “AutoFill Passwords”
5. Check “Bitwarden”
6. You can uncheck “iCloud Keychain” if you want

Step 9: Enable Face ID / Touch ID Makes unlocking Bitwarden super fast.

1. Open Bitwarden app
2. Tap Settings (gear icon)
3. Tap “Security”
4. Enable “Unlock with Face ID” (or Touch ID)
5. Verify with your face/fingerprint

Done! Bitwarden is ready on your iPhone.

Quick Test:

1. Open Safari on your iPhone

2. Go to any login page

3. Tap the password field

4. You should see a key icon above the keyboard

5. (You won’t have passwords yet, but the option should be there)

For Android

Step 1: Open the Google Play Store

Step 2: Search for "Bitwarden"

Step 3: Look for "Bitwarden Password Manager" - By "8bit Solutions LLC"

Step 4: Tap "Install"

Step 5: Open Bitwarden

Step 6: Log In - Email - Master password - 2FA code

Step 7: Enable Autofill Service

1. Open Android Settings (not Bitwarden)
2. Search for "Autofill"
3. Tap "Autofill service"
4. Select "Bitwarden"
5. Grant permissions when asked

Step 8: Enable Biometric Unlock

1. Open Bitwarden app
2. Tap Settings (☰ menu → Settings)
3. Tap "Security"
4. Enable "Unlock with Biometrics"
5. Verify with fingerprint/face

Step 9: Disable Battery Optimization This ensures Bitwarden works smoothly.

1. Android Settings → Apps

2. Find Bitwarden
3. Tap Battery
4. Select "Don't optimize"

Done! Bitwarden is ready on Android.

Quick Test: 1. Open any app with a login (like Gmail) 2. Tap a password field 3. You should see "Autofill with Bitwarden" suggestion 4. (Nothing to fill yet, but the option should appear)

What You've Accomplished

☑ Installed Bitwarden in your web browser

☑ Tested that the extension works

☑ Installed Bitwarden on your phone

☑ Enabled auto-fill on your phone

☑ Set up biometric unlock (fingerprint/face)

☑ Tested that everything's working

Bitwarden is now installed everywhere you need it.

Using Bitwarden Daily (Preview)

Now that it's installed, here's what daily use looks like:

On your computer:

1. Visit a website

2. Click in the password field

3. Press `Ctrl+Shift+L` (or click Bitwarden icon)

4. Password fills automatically

5. Click login

On your phone:

1. Tap a password field

2. Tap the Bitwarden suggestion

3. Use Face ID/fingerprint

4. Password fills

5. Tap login

It's actually faster than what you're doing now.

What's Next

In Chapter 4, we'll start adding your passwords to Bitwarden:

- How to save passwords manually

- How to import all your existing passwords from your browser

- How to generate strong passwords for new accounts

Then in Chapter 5, we'll practice using Bitwarden every day until it's second nature.

Take a quick break. When you're ready, let's fill up that vault with your passwords.

Quick Troubleshooting

Browser extension won't install:

- Try a different browser

- Check that you're using a recent browser version

- Restart your browser and try again

Can't log in on mobile:

- Make sure you're typing master password correctly (auto-correct can mess it up)

- Make sure 2FA code hasn't expired (they change every 30 seconds)

- Try typing manually instead of copy/paste

Auto-fill not working on phone:

- iPhone: Check Settings → Passwords → Bitwarden is checked

- Android: Check Settings → Autofill Service → Bitwarden selected

- Restart your phone

Everything else:

- Log out and log back in

- Close the app completely and reopen

- Uninstall and reinstall (your passwords are safely stored on Bitwarden's servers)

Still stuck? Jump to Chapter 9 (Common Problems) for more help.

Ready for Chapter 4? Let's add some passwords!

Chapter 4: Adding Your Passwords

Three Ways to Fill Your Vault

1. **Save passwords as you use them** (easiest, happens automatically)
2. **Manually add important passwords** (takes 2 minutes each)
3. **Import all at once from your browser** (fastest for lots of passwords)

We'll cover all three, starting with the easiest.

Method 1: Save As You Go (Automatic)

This is how you'll add most passwords going forward.

Here's what happens:

Step 1: You log into a website

- Enter your username

- Enter your password

- Click "Log In"

Step 2: Bitwarden notices A notification pops up: "Do you want to save this password?"

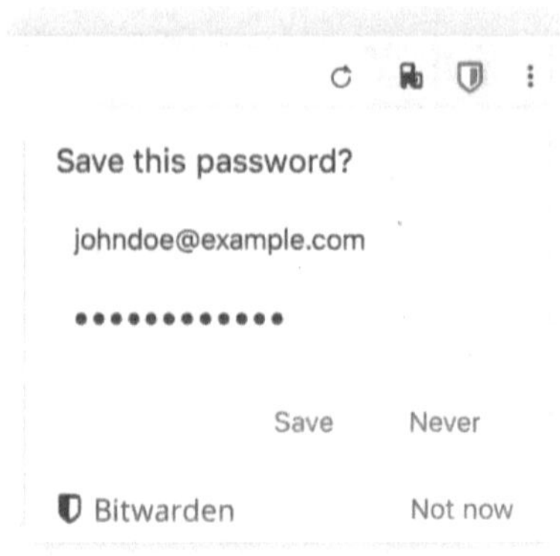

Figure 8 - Set Password - AI Generated

Step 3: You click "Yes" Bitwarden saves it automatically.

That's it. Next time you visit that site, Bitwarden will fill it in for you.

Let's try it:

1. Go to any website where you have an account (Twitter, Reddit, etc.)
2. Log in like normal
3. Watch for the Bitwarden notification
4. Click "Save" when prompted
5. Open Bitwarden (click the icon)
6. You should now see that login saved

If the notification doesn't appear:

- Click the Bitwarden icon

- Click "Add Item"

- Fill in the website, username, and password manually

- Click "Save"

Method 2: Add Passwords Manually

For important accounts, let's add them properly right now.

Let's add your email account:

Step 1: Open Bitwarden Click the Bitwarden icon in your browser.

Step 2: Click "Add Item" The plus (+) button at the top.

Step 3: Fill in the Details

Name: What you'll see in your list

- Example: "Gmail - Personal"

Username: Your email address or username

- Example: "yourname@gmail.com"

Password: Your current password for that site

- Just type it in for now

Website (URI): The login page

- Example: "https://mail.google.com"

- Or just "gmail.com"

- Bitwarden figures it out

Step 4: Click "Save"

Step 5: Test It

1. Go to Gmail

2. Click in the username field

3. Press `Ctrl+Shift+L` (or click Bitwarden and select your login)

4. It should fill in your username and password

5. Click "Sign In"

It works!

Now add a few more important accounts:

- Your main email

- Your bank

- Amazon/shopping sites

- Social media

Just repeat the same process for each one.

Method 3: Import from Your Browser

If you have dozens of passwords saved in Chrome, Firefox, or another browser, you can import them all at once.

Importing from Chrome

Step 1: Export from Chrome

1. Open Chrome
2. Type in address bar: `chrome://settings/passwords`
3. Click the three dots (⋮) next to "Saved Passwords"

4. Click “Export passwords”
5. Confirm with your computer password
6. Save the file somewhere you’ll remember (like your Desktop)
7. Name it something like “chrome-passwords.csv”

IMPORTANT: This file contains all your passwords in plain text. We’ll delete it as soon as we’re done importing.

Step 2: Import to Bitwarden

1. Go to vault.bitwarden.com (log in if needed)
2. Click “Tools” in the top menu
3. Click “Import Data”
4. Under “1. Select the format,” choose “Chrome (csv)”
5. Click “Choose File”
6. Select the file you saved (chrome-passwords.csv)
7. Click “Import Data”

Step 3: Verify

1. Click “My Vault”
2. You should see all your passwords listed
3. Spot-check a few to make sure they imported correctly

Step 4: DELETE THE EXPORT FILE

This is critical:

1. Find that chrome-passwords.csv file

2. Delete it

3. Empty your Recycle Bin / Trash

4. It contained all your passwords unencrypted!

Done! All your Chrome passwords are now safely in Bitwarden.

Importing from Firefox

Step 1: Export from Firefox

1. Open Firefox
2. Type in address bar: `about:logins`
3. Click the three dots (···) at the top right
4. Click "Export Logins"
5. Acknowledge the security warning
6. Save as "firefox-passwords.csv"

Step 2: Import to Bitwarden

1. vault.bitwarden.com → Tools → Import Data
2. Format: "Firefox (csv)"
3. Choose your file
4. Import

Step 3: Verify and Delete

Check your vault, then DELETE the export file.

Importing from Edge or Safari

Same process:

1. Export from your browser (Google how to export passwords from your specific browser)

2. Import to Bitwarden (select the right format)

3. Verify it worked

4. DELETE the export file

Using the Password Generator

When you create a new account on a website, let Bitwarden create a super-strong password for you.

Here's how:

Step 1: You're signing up for a new website You've filled in your username and email. Now you need a password.

Step 2: Click the Bitwarden Icon

Step 3: Click the "Generator" Tab (It's next to "My Vault")

Figure 9 – Password Generator – LP/PLP

Step 4: Click "Generate" Bitwarden creates something like: `X7$mK9#pL2@nQ5*j`

Step 5: Click the Copy Icon (Or just click the password itself)

Step 6: Paste it into the Website Paste in both the "Password" and "Confirm Password" fields.

Step 7: Complete the Signup

Step 8: Save to Bitwarden When Bitwarden asks "Save this password?" click "Yes."

Why this is great:

- Super strong password (would take billions of years to crack)

- Unique to this website

- You never have to remember it

- You never even have to see it again

From now on: Every new account = click Generate = paste = done.

Organizing Your Passwords (Basic)

You don't need fancy organization yet, but here are two simple tricks:

Use Favorites for Important Stuff

What to favorite:

- Your email accounts

- Your bank

- Accounts you use daily

How to favorite:

1. Open the password in Bitwarden

2. Click the star icon

3. Now it appears at the top of your list

Give Things Clear Names

Instead of “Login” or “Amazon” use:

- “Gmail - Personal”

- “Bank of America

- Checking”

- “Amazon - Main Account”

This helps when you have multiple accounts at the same place.

To rename:

1. Click the password in Bitwarden
2. Click “Edit” (pencil icon)
3. Change the “Name” field
4. Click “Save”

What You’ve Accomplished

☑ Saved passwords automatically while browsing

☑ Manually added important accounts

☑ Imported all your browser passwords (optional)

☑ Learned to use the password generator

☑ Organized with favorites and clear names

☑ Deleted the exported password file (critical!)

Your vault now has all your passwords.

Common Questions

Q: How many passwords should I have in Bitwarden?

A: However many accounts you have! Most people end up with 50-100. Some have 200+. There's no limit.

Q: Should I delete passwords from my browser now?

A: Eventually, yes. But don't rush it. Use Bitwarden for a week first to make sure you're comfortable. Then you can clear your browser's saved passwords.

Q: What if I made a mistake importing?

A: You can delete passwords individually (Edit → Delete) or delete everything and start over.

Q: Can I add passwords on my phone?

A: Yes! The process is similar—tap the + button in the Bitwarden app. But it's easier to add most passwords on your computer.

Q: What about that exported CSV file?

A: It should be DELETED. If you didn't already, do it now. It's a security risk.

What's Next

In Chapter 5, we'll practice actually USING Bitwarden every day:

- Logging into websites with one click

- Creating new accounts with generated passwords

- Getting fast with keyboard shortcuts

- Building the habit

This is where Bitwarden goes from “thing I’m trying” to “how did I ever live without this.”

Ready? Let’s practice in Chapter 5.

Chapter 5: Using Bitwarden Every Day

Your New Daily Routine

For the next week, you're going to practice using Bitwarden until it becomes automatic. By the end of the week, you won't remember how you managed passwords before.

Here are the three things you'll do most often:

1. **Log into websites** (using saved passwords)
2. **Create new accounts** (with generated passwords)
3. **Change passwords** (when needed)

Let's practice each one.

Logging Into Websites (The Easy Way)

You visit a website, need to log in. Here's what you do:

The One-Key Method (Fastest)

1. **Go to the login page**
   - Example: mail.google.com
2. **Click in the username field**
   - Just click once
3. **Press `Ctrl + Shift + L`**
   - On Mac: `Cmd + Shift + L`
   - Bitwarden fills everything in
4. **Click "Log In"**
   - Done in 3 seconds

This is the method to master. It's faster than typing and works everywhere.

The Click Method (Also Easy)

If you forget the keyboard shortcut:

Go to the login page

5. **Click the Bitwarden icon** in your toolbar
6. **Click your login** from the list
7. **Fields fill automatically**
8. **Click "Log In"**

On Your Phone

iPhone/iPad:

1. Tap the username or password field
2. Tap the **key icon** above the keyboard
3. Use Face ID / Touch ID
4. Tap your login
5. Done

Android:

1. Tap a login field

2. Tap the Bitwarden suggestion

3. Use fingerprint / face

4. Tap your login

5. Done

Practice Exercise: Log Into 5 Websites

Right now, practice this:

1. Pick 5 websites where you have accounts
2. Log out of each one
3. Log back in using Bitwarden
4. Use the `Ctrl + Shift + L` method each time

Websites to practice with:

- Your email

- Amazon or shopping site

- Social media (Facebook, Twitter, Instagram)

- Banking (if you're comfortable)

- Any forum or website

After doing this 5 times, it'll feel natural.

Creating New Accounts (The Right Way)

From now on, every new account you create should have a strong, unique password generated by Bitwarden.

Here's the workflow:

Step 1: You're Signing Up

You're on a website, creating a new account. You've entered your email and username.

Step 2: Generate Password FIRST

Before filling the password field:

1. Click Bitwarden icon
2. Click "Generator" tab
3. Click "Generate" (or just click the password—it copies)
4. Leave this window open

Step 3: Paste Into Website

1. Click the password field on the website
2. Paste (`Ctrl + V` or right-click → Paste)
3. Do the same for "Confirm Password" field

Step 4: Complete Signup

Fill in any other required fields and submit.

Step 5: Bitwarden Asks to Save

"Do you want to save this password?"

Click "Yes"

Step 6: Verify

1. Open Bitwarden
2. Find the new login
3. Make sure it saved correctly

That's it. You now have a super-strong password you'll never need to remember or type again.

Practice Exercise: Create a Test Account

Let's practice with a real example:

1. Go to **reddit.com** (or another site you don't have an account on)
2. Click "Sign Up"
3. Use Bitwarden generator for the password
4. Complete the signup
5. Save to Bitwarden
6. Log out
7. Log back in using Bitwarden (`Ctrl + Shift + L`)

This workflow should become automatic:

- New account = Generate password = Save = Never think about it again

Changing Passwords (When Needed)

Sometimes you need to change a password:

- Website got hacked

- You used a weak password before

- You reused a password (fixing that now)

- Website requires periodic changes

Here's how:

Step 1: Log Into the Website

Use Bitwarden to fill your current password.

Step 2: Find "Change Password"

Usually in:

- Settings

- Security

- Account Settings

- Profile

Step 3: Enter Current Password

Click Bitwarden, copy your current password, paste it.

Step 4: Generate New Password

1. Click in the "New Password" field
2. Open Bitwarden
3. Go to Generator
4. Generate and copy new password
5. Paste in both "New Password" and "Confirm Password"

Step 5: Save the Change

Click "Change Password" or "Update" on the website.

Step 6: Update Bitwarden

Bitwarden usually asks: "Update this password?"

Click "Update"

If it doesn't ask:

1. Open Bitwarden

2. Find that login

3. Click Edit

4. Update the password field 5. Save

Step 7: Test It

1. Log out
2. Log back in
3. Verify the new password works

Building the Habit (Week 1 Challenge)

For the next 7 days:

Day 1-2: Log into 10 websites using Bitwarden

- Practice the keyboard shortcut

- Get comfortable with the workflow

- Note which sites work perfectly, which need manual help

Day 3-4: Change 5 weak passwords

- Look for accounts where you used "password123" or similar

- Generate strong new passwords

- Update Bitwarden

Day 5-6: Create 2-3 new accounts

- Sign up for something (forum, news site, anything)

- Use the generator every time

- Practice the workflow

Day 7: Review your vault

- Look at everything you've added

- Add any missing accounts

- Star your favorites

- Celebrate how much more secure you are

By Day 8: You won't want to go back to the old way.

Keyboard Shortcuts to Memorize

There are only THREE you need to know:

1. `Ctrl + Shift + L` - Auto-fill login
   - This is the one you'll use 20 times a day
   - Learn it, love it, live it
2. `Ctrl + Shift + Y` - Open Bitwarden
   - Quick access to your vault
   - Faster than clicking the icon
3. `Ctrl + Shift + 9` - Generate password
   - Quick password generation
   - Optional (can also use the generator tab)

Just the first two will change your life.

Tips for Success

Tip 1: Trust the System

Your brain will say "I should remember this password just in case."

You don't need to. Bitwarden remembers it. Let it do its job.

Tip 2: Use It for Everything

The more you use Bitwarden, the more natural it becomes. Use it for:

- Important sites (banks, email)

- Unimportant sites (random forums)

- Everything in between

Tip 3: Don't Delete Browser Passwords Yet

Keep them for one week while you learn Bitwarden. Then delete them.

Tip 4: It's Okay to Look Things Up

Forgot the keyboard shortcut? Click the icon instead. Forgot how to generate a password? Read this chapter again. You'll remember with practice.

Tip 5: Fix Problems Immediately

If Bitwarden doesn't fill a password correctly, fix it right then:

- Edit the login

- Update the website URL

- Test it again

- Don't ignore problems

Common Situations

"Bitwarden suggested the wrong password"

You have multiple accounts for this site. Click Bitwarden icon, choose the right one.

"Nothing happens when I press Ctrl+Shift+L"

Make sure:

- You're clicked in a password field

- You're on a login page

- Bitwarden is unlocked

- You have a login saved for this site

"The password didn't work"

- Did you update it on the website but not in Bitwarden?
- Are you using the right account?
- Try clicking "Forgot Password" and reset it, then update Bitwarden

"I created an account but Bitwarden didn't ask to save"

Manually add it:

1. Bitwarden → Add Item

2. Fill in the details

3. Save

What You've Learned

☑ How to log in with one keystroke

☑ How to generate strong passwords for new accounts

☑ How to change passwords and update Bitwarden

☑ The three essential keyboard shortcuts

☑ A week-long plan to build the habit

What's Next

In Chapter 6, we'll organize your vault:

- Using folders to keep things tidy

- Adding notes to passwords (account numbers, security questions)

- Dealing with different types of information (credit cards, etc.)

But first: **Practice what you learned in this chapter.**

Spend a few days using Bitwarden for real. Get comfortable. Build the muscle memory.

When you're ready, Chapter 6 will help you organize everything perfectly.

Remember: The best password manager is the one you actually use. Make using Bitwarden a daily habit, and you're already 90% more secure than most people.

You've got this!

Chapter 6: Keeping Things Organized

You Don't Need Complicated Organization

Good news: You don't need a perfect filing system. Bitwarden has excellent search, so you can find anything in seconds.

But a little organization makes life easier. This chapter shows you the simple tricks that actually help.

Using Folders (The Simple Way)

Folders are like drawers in a filing cabinet. You don't need many—just enough to keep things tidy.

Creating Your First Folder

Step 1: Log into vault.bitwarden.com

Step 2: Click "Settings" (bottom left)

Step 3: Click "Folders"

Step 4: Click "Add Folder"

Step 5: Name it something simple

- "Banking"

- "Shopping"

- "Work"

- "Important"

Step 6: Click Save

Done! You have a folder.

My Recommended Folder Setup

Start with just **3-5 folders:**

```
🗀 Important (email, banking)
🗀 Shopping (Amazon, eBay, etc.)
🗀 Social Media
🗀 Work
🗀 Everything Else (or no folder)
```

That's it. Don't overthink it.

Moving Passwords Into Folders

Step 1: Open the password you want to organize

Step 2: Click "Edit" (pencil icon)

Step 3: Under "Folder," select from dropdown

Step 4: Click "Save"

Takes 5 seconds per password.

Or leave most things unfiled. Search works great. Only organize what helps YOU find things faster.

Using Favorites (Even Simpler)

Favorites are for stuff you use all the time.

What to Favorite

Star these:

- Your email(s)

- Your bank

- Accounts you use daily

- "I can't live without this" accounts

Don't star:

- That forum you visited once in 2019

- Accounts you use monthly

- Everything (defeats the purpose)

How to Favorite

Click the star next to any password. That's it.

Favorites appear at the top of your vault, so they're always easy to find.

Adding Notes to Passwords

The Notes field is perfect for extra information you might need.

What to Put in Notes

For banking:

```
Account Number: 123456789
Customer Service: 1-800-123-4567
Account opened: January 2020
```

For websites:

```
Security Questions:
Q: Mother's maiden name
A: RandomWord123 (fake answer!)
```

```
Q: First pet
A: AnotherRandom456 (fake answer!)
```

For subscriptions:

```
Plan: Premium Annual
Cost: $99/year
Renewal: January 15
Cancel before: December 15 to avoid renewal
```

How to add notes:

1. Edit the password
2. Scroll to "Notes" field
3. Type your information
4. Save

Pro tip: Include searchable keywords so you can find it later.

Storing Credit Cards

Bitwarden can also store credit card information.

Adding a Credit Card

Step 1: Click "Add Item"

Step 2: Change "Type" from "Login" to "Card"

Step 3: Fill in the details:

- Cardholder Name: Your name as it appears on card

- Card Number: Your full card number

- Brand: Visa, Mastercard, etc. (auto-detects)

- Expiration Month/Year

- Security Code (CVV): The 3 digits on back

Step 4: Name it clearly

- "Chase Visa- Primary"

- "AmEx- Business"

Step 5: Add to folder (optional) - "Banking" or "Payment Methods"

Step 6: Save

Using Stored Credit Cards

On computer:

1. Click Bitwarden icon

2. Click "Cards" tab

3. Click the card you want

4. Copy the number, date, CVV as needed

On phone: Similar—open Bitwarden app, go to Cards, copy what you need.

Is this safe?

Yes. Your credit card info is encrypted the same way as your passwords. Only you can see it.

Storing Other Information

Secure Notes

For anything that doesn't fit elsewhere:

1. Add Item

2. Type: “Secure Note”
3. Name it (“WiFi Password,” “Alarm Code,” etc.)
4. Put information in the Notes field
5. Save

Good for:

- WiFi passwords

- Alarm codes

- Software license keys

- Important confirmation numbers

- Emergency contacts

Identity Information (Optional)

Bitwarden can store address and personal info for form auto-fill.

To add:

1. Add Item

2. Type: “Identity”

3. Fill in your information 4. Save

Useful for:

- Shipping addresses

- Contact information you type repeatedly

- Multiple addresses (home, work, PO box)

Most people don't bother with this. It's there if you want it.

Search: Your Best Friend

Honestly, you can skip most organization if you just use search.

How to search:

On computer:

1. Click Bitwarden icon
2. Start typing in the search box
3. Results appear instantly

Or: Press `Ctrl+Shift+Y`, then start typing.

On phone: Open Bitwarden app, tap the search icon, type.

Search finds:

- Password names
- Usernames
- Website URLs
- Anything in the notes field

Example: Type "bank" and all your banking-related passwords appear.

This is often faster than browsing folders.

Cleaning Up Your Vault

Every few months, do a quick cleanup:

Delete Unused Accounts

Found a password for:

- Website that doesn't exist anymore
- Account you closed
- Duplicate entry
- Random site you'll never use again

Delete it:

1. Open the item
2. Click Edit
3. Scroll to bottom
4. Click "Delete"
5. Confirm

Update Outdated Information

If you:

- Changed your email address
- Got a new credit card
- Updated phone number

Update those in Bitwarden too.

Fix Confusing Names

Rename unclear entries:

- "Login" → "Netflix - Family Account"

- "Amazon" → "Amazon - Personal Shopping"

Makes searching easier

What You've Learned

☑ Create simple folders (3-5 is enough)

☑ Use favorites for daily-use accounts

☑ Add helpful notes to passwords

☑ Store credit cards securely

☑ Use secure notes for misc. information

☑ Search is your friend (often better than organizing)

☑ Clean up your vault occasionally

Common Questions

Q: How many folders should I have?

A: 3-5 is perfect. More than 10 is probably overkill.

Q: What if I can't decide which folder something goes in?

A: Then don't put it in a folder. Use search instead.

Q: Should I store my Social Security Number in Bitwarden?

A: That's your call. It's encrypted and secure, but some people prefer to keep extremely sensitive docs in a physical safe.

Q: Can I change my folder structure later?

A: Yes! Move things around anytime. It's flexible.

Q: Is there a limit to how much I can put in Notes?

A: Technically yes (10,000 characters), but you'll never hit it with normal use.

What's Next

In Chapter 7, we'll cover sharing passwords with family members—like sharing the Netflix password or giving your spouse access to the bank account.

If you don't need to share passwords with anyone, you can skip Chapter 7 and jump straight to Chapter 8 (Staying Secure).

But for now, take 10 minutes and organize what you have: - Create 3 folders - Star your favorites - Add notes to your important accounts - Delete anything obviously useless

Your vault is now organized and easier to navigate. Nice work!

Chapter 7: Sharing Passwords with Family

The Family Password Problem

You probably share passwords with family:

- Netflix, Disney+, streaming services
- Utilities (electric, water, internet)
- Shared bank accounts
- Family calendar/email

How you probably share them now:

- Text message
- Sticky note on fridge
- "Hey honey, what's the WiFi password again?"
- Everyone uses the same password for everything

There's a better way.

Bitwarden Families: The Solution

What it is: A shared vault that up to 6 family members can access—separate from your personal vault.

How it works:

- You keep your private passwords private
- Shared accounts (Netflix, etc.) go in the family vault
- Everyone in the family can access shared passwords

- You control who sees what

Cost: $40 per year (that's $6.67 per person)

Worth it? Absolutely. Your family stays secure, and you stop texting passwords.

Setting Up Families Organization

Step 1: Subscribe

1. Log into vault.bitwarden.com
2. Click "Settings" (bottom left)
3. Click "Organizations"
4. Click "New Organization"
5. Select "Families" plan
6. Enter organization name (e.g., "Smith Family")
7. Enter your email for billing
8. Add payment method
9. Complete purchase

Done! You now have a Families organization.

Step 2: Invite Family Members

Who to invite:

- Spouse/partner

- Kids (if old enough to use Bitwarden responsibly)

- Other family members who need access

- Up to 6 people total (including you)

How to invite:

1. Organizations → [Your Family Name]

2. Click “Manage”
3. Click “Invite User”
4. Enter their email address
5. Leave access as “User” (default)
6. Click “Save”

They’ll receive an email:

- Click the invitation link

- Create a Bitwarden account (if they don’t have one)

- Accept the invitation

- They’re now in the family organization

Repeat for each family member.

Step 3: Create Collections

Collections are like folders, but for the family vault.

Create these collections:

For everyone:

- “Streaming Services” (Netflix, Disney+, etc.)

- “WiFi & Internet”

- “Shared Shopping” (Amazon family account)

For adults only:

- “Banking & Bills”

- “Important Accounts”

How to create:

1. Organizations → [Family Name] → Manage

2. Click “Collections” tab
3. Click “New Collection”
4. Name it (e.g., “Streaming Services”)
5. Save

Create 3-5 collections to start. You can always add more later.

Step 4: Assign Collection Access

Now control who can see what.

Example: “Streaming Services”

- All family members can access

- Kids included

Example: “Banking & Bills”

- Only you and spouse

- Kids excluded

How to set access:

1. Organizations → Manage → Collections
2. Click the collection name
3. Under “Access,” select which family members
4. Choose their permission:
   – “Can View” - Can see and use passwords (safest for kids)
   – “Can Edit” - Can modify passwords
5. Save

Recommendation:

- Give spouse "Can Edit" on everything

- Give kids "Can View" on appropriate collections

Sharing Your First Password

Let's share the Netflix password with the family.

Step 1: Move It to Organization

If you already have Netflix in your personal vault:

1. Open that password
2. Click "Edit"
3. Scroll to "Ownership"
4. Change from "Personal Vault" to "[Family Name]"
5. Select collection: "Streaming Services"
6. Save

If you need to add it fresh:

1. Click "Add Item"
2. Under "Ownership," select your family organization
3. Select "Streaming Services" collection
4. Fill in Netflix login details
5. Save

Step 2: Everyone Can Now Access It

Family members:

1. Open their Bitwarden

2. Click "Organizations" or "Collections"

3. Open “Streaming Services”

4. See Netflix password

5. Use it to log in

That’s it! No more texting passwords.

What to Share vs. Keep Private

Share These:

Household Services:

- Streaming (Netflix, Disney+, Hulu)

- Utilities (electric company, water, internet)

- Shared shopping accounts

- Family subscriptions

Adults Only:

- Bank accounts

- Credit cards

- Bills and finances

- Insurance portals

Keep Private:

Your personal accounts:

- Your personal email

- Your personal social media
- Your work accounts

- Your personal shopping

- Your Amazon wish list (keep surprises secret!)

Rule of thumb: If it's just yours, keep it in your personal vault. If the family needs it, share it.

Teaching Family Members

Not everyone is tech-savvy. Here's how to help:

For Your Spouse/Partner

Show them:

1. How to open Bitwarden

2. Where to find shared passwords (Organizations tab)

3. How to copy a password

4. One practice login together

Give them this book (or the relevant chapters).

For Kids

Start simple:

1. Install Bitwarden on their phone

2. Show them where streaming passwords are

3. Demonstrate once

4. Supervise first few times

5. Give them "View Only" access (they can't change or delete)

Don't overwhelm them. They just need to know: "Open Bitwarden, find Netflix, copy password."

For Less Tech-Savvy Family

Be patient:

- Write down simple instructions

- Make a cheat sheet

- Expect to help a few times

- Eventually they'll get it

Consider: For elderly parents, you might just give them the passwords directly instead of making them learn Bitwarden.

Managing Shared Accounts

When Someone Changes a Password

If you change Netflix password:

1. Change it on Netflix's website
2. Update it in Bitwarden (in the family organization)
3. Tell family: "Updated Netflix password—it's in Bitwarden"
4. Everyone automatically has the new password

No more group texts with the new password!

When Someone Leaves

Kid going to college? They keep their Bitwarden account but you can:

1. Remove them from family collections

2. Or keep them in certain collections

Divorce/separation?

1. Remove them from organization

2. Change all shared passwords immediately

3. Create new family organization if needed

How to remove someone:

1. Organizations → Manage → Members

2. Find their name

3. Click gear icon → Remove

4. Confirm

Common Scenarios

Scenario: Babysitter Needs WiFi

Don't add them to family organization. Instead:

1. Create a guest WiFi network (if your router supports it)

2. Give them that password directly

3. Or use Bitwarden Send (temporary sharing - see sidebar)

Scenario: Split Streaming Costs

Friend pays half of Netflix?

1. Don't add them to family organization

2. Share that ONE password with Bitwarden Send

3. Or give it to them directly

Scenario: Multiple Kids, Different Ages

Teenager and young kid?

- Teenager: Access to streaming, appropriate sites

- Young kid: Maybe just streaming, supervised

- Create different collections for different ages

Scenario: Want to Check Who Accessed What

Bitwarden Families doesn't have detailed access logs for free tier. You'll just have to trust your family (which you probably should anyway).

Benefits You'll Notice

Week 1: - Stop texting passwords - Everyone can log into Netflix on their own

Month 1:

- Spouse stops asking "What's the bank password?"

- Kids can connect to WiFi without asking

Year 1:

- When a website gets hacked, change one password in Bitwarden

- Everyone automatically has the new password

- No group chat needed

It just works, silently, in the background.

What You've Learned

- ☑ Set up Bitwarden Families organization
- ☑ Invited family members
- ☑ Created collections for shared passwords
- ☑ Shared your first passwords
- ☑ Know what to share vs. keep private
- ☑ Can manage family access

Common Questions

Q: Does Families cost extra?

A: Yes, $40/year. But it includes premium features for all 6 family members (normally $10 each), so it's actually a deal.

Q: Can I try before buying?

A: Set up the free organization first, then upgrade when you're ready. Or just subscribe—it's less than $4/month for the whole family.

Q: What if I only have 2 people?

A: Families still works great for couples. You get 6 seats, you just won't use them all.

Q: Can my kid see my personal passwords?

A: No. Only what's in the shared collections. Your personal vault is completely separate and private.

Q: What happens if I cancel Families?

A: Shared passwords go back to personal vaults. You keep everything, just lose the sharing capability.

Q: Do family members need their own accounts?

A: Yes, each person has their own Bitwarden account with their own master password. This is more secure than sharing one account.

What's Next

In Chapter 8, we'll talk about staying secure long-term:

- Monthly security checks

- What to do when websites get hacked

- Backing up your passwords

- Simple habits that keep you safe

Whether you set up family sharing or not, Chapter 8 is important for everyone.

Take a break if you need one. When you're ready, let's make sure you stay secure!

Chapter 8: Staying Secure

Security Isn't "Set It and Forget It"

You've done the hard work—Bitwarden is set up, your passwords are strong, you're using it daily. Great!

But security needs occasional attention. Think of it like changing your car's oil—do it regularly, and everything runs smoothly.

Good news: It's easy and doesn't take much time.

Your Monthly 15-Minute Security Check

Once a month, spend 15 minutes on security. That's it.

Pick a recurring time:

- First Sunday of each month

- When you pay bills

- Your birthday (easy to remember)

Set a calendar reminder right now. Seriously, do it.

The 15-Minute Routine

Month 1: Check for Weak Passwords

If you have Premium ($10/year):

1. vault.bitwarden.com

2. Tools → Vault Health Reports

3. Click "Weak Passwords Report"

4. Change the worst 5-10 passwords

If you have Free:

1. Look through your vault
2. Find passwords like "password123" or "fluffy2024"
3. Change them to generated passwords

Month 2: Check for Reused Passwords

Premium:

1. Tools → Vault Health Reports
2. "Reused Passwords Report"
3. Change passwords you're using on multiple sites

Free:

1. Think about passwords you reused
2. Look for accounts where you used the same password
3. Change them

Month 3: Check for Breaches

1. Visit **haveibeenpwned.com**
2. Enter your email address
3. See if you've been in any breaches
4. If yes, change those passwords

Repeat this cycle every month.

That's it. 15 minutes per month keeps you secure.

When a Website Gets Hacked (It Happens)

You'll hear news: "XYZ Company hacked, 10 million accounts compromised."

Don't panic. Here's what to do:

Step 1: Check If You're Affected

Visit haveibeenpwned.com

- Enter your email

- See if you show up in that breach

Or just assume you're affected and take action anyway.

Step 2: Change the Password Immediately

On the compromised website:

1. Log in
2. Go to Settings → Security
3. Change password
4. Use Bitwarden generator for new password

In Bitwarden:

1. Find that login
2. Edit it
3. Update with new password
4. Save

Step 3: Check for Reuse

Critical: Did you use that same password anywhere else?

If yes:

1. Change it everywhere immediately

2. Every site that used that password needs a new one

3. Use Bitwarden generator—make each one unique

Step 4: Enable 2FA (If You Haven't)

If the site offers two-factor authentication:

1. Enable it

2. Use authenticator app (not SMS if possible)

3. Save recovery codes in Bitwarden Notes field

Step 5: Monitor Your Accounts

For the next month:

- Watch for suspicious activity

- Check bank statements

- Review account access logs

- Be alert for phishing emails

Most breaches don't affect you directly if you:

- Had a unique password (only that site affected)

- Change it quickly - Have 2FA enabled

Backing Up Your Passwords

What if:

- Bitwarden goes down (unlikely but possible)

- You forget your master password (and your backup)

- Your account gets locked somehow

Solution: Keep a backup.

How to Export Your Vault

Every 3 months:

1. vault.bitwarden.com
2. Tools → Export Vault
3. Select "JSON" format
4. Enter your master password
5. Click "Export Vault"
6. Save the file

Name it: `bitwarden-backup-2024-03.json`

Where to Store the Backup

Option 1: USB Drive (Recommended)

1. Save export to encrypted USB drive

2. Store USB in safe or locked drawer

3. Update every 3 months

Option 2: Encrypted File on Computer

1. Encrypt the export file (7-Zip, WinRAR, etc.)

2. Use a strong encryption password

3. Store on your computer

4. Also copy to external drive

Option 3: Printed Backup (Emergency Only) For your absolute most critical passwords:

1. Print them on paper

2. Store in home safe

3. Include: email, bank, Bitwarden master password

4. Update when they change

CRITICAL: The export file contains ALL your passwords in readable form. Treat it like gold. Encrypt it or lock it away physically.

Delete Old Exports

After you've created a new backup:

1. Delete the old export file

2. Empty your trash/recycle bin

3. Keep only the most recent backup

Simple Daily Habits

Small habits make a big difference:

☑ Always Use the Generator

Creating new account? Generate password. Every time. No exceptions.

☑ **Save Immediately**

Changed a password on a website? Update Bitwarden right away. Don't wait.

☑ **Never Text Passwords**

Use Bitwarden sharing or family organization. Not text/email.

☑ **Lock When Away**

Leaving your computer? Lock it (`Windows + L` or `Cmd + Control + Q`). Bitwarden locks with it.

☑ **Check URLs Before Entering Passwords**

Make sure you're on the real website, not a phishing site. Bitwarden helps—it won't auto-fill on fake sites.

What to Do If…

You Lose Your Phone

Your 2FA authenticator is on it!

Don't panic:

1. Use your 2FA recovery code (you saved it, right?)
2. Log into Bitwarden on computer
3. Set up 2FA on new phone
4. Generate new recovery codes
5. Deauthorize old phone if possible

Prevention: Save 2FA recovery codes with your master password backup.

You Forget Your Master Password

First: Check where you wrote it down (safe, drawer, safe deposit box).

Second: Try the password hint (if you set one).

Third: Try variations you might have used.

Last Resort: If you truly can't remember it and have no backup:

1. You're locked out forever
2. Create new Bitwarden account
3. Import from your backup export (if you have one)
4. Start over (if you don't)

This is why writing it down in a safe place is critical.

Someone Gains Access to Your Account

If you suspect compromise:

1. **Change master password immediately**
   - Settings → Change Master Password
2. **Reset 2FA**
   - Disable old 2FA
   - Set up new 2FA
   - New recovery codes
3. **Deauthorize all sessions**
   - Settings → My Account → Deauthorize Sessions

    - Forces logout everywhere
4. **Check for unauthorized changes**
    - Review vault for new/changed items
    - Look for suspicious activity
5. **Change critical passwords**
    - Email
    - Banking
    - Anywhere sensitive
6. **Report to Bitwarden** if needed

Security Best Practices (Quick Reference)

☑ Do These:

- Use unique, generated passwords for every site

- Enable 2FA everywhere (Bitwarden + important sites)

- Keep software updated (Bitwarden, browser, OS)

- Use strong master password (passphrase method)

- Back up quarterly

- Monthly security check

- Lock computer when away

- Check URLs before entering passwords

✗ Don't Do These:

- Reuse passwords

- Use weak passwords

- Share master password

- Store passwords unencrypted

- Ignore breach notifications

- Skip 2FA

- Use public WiFi without VPN (for sensitive stuff)

- Click suspicious email links

The Reality Check

You don't need to be paranoid.

Following these basics puts you ahead of 95% of internet users:

1. Password manager with strong master password

2. Two-factor authentication

3. Unique passwords everywhere

4. Monthly check-ins

5. Quarterly backups

That's it. You're doing great.

Staying Current

Bitwarden improves constantly:

- New features

- Security updates

- Bug fixes

How to stay updated:

- Update browser extension when prompted

- Update mobile apps from app store

- Read Bitwarden blog occasionally (bitwarden.com/blog)

- Follow @bitwarden on Twitter for major announcements

Don't stress about every update. Just keep things reasonably current.

What You've Learned

☑ Monthly 15-minute security routine

☑ How to respond to breaches

☑ Backing up your vault

☑ Simple daily security habits

☑ Emergency procedures (lost phone, forgotten password)

☑ Security best practices

☑ What to worry about (and what not to)

What's Next

Chapter 9 (the last chapter!) covers common problems and how to fix them:

- Can't log in

- Auto-fill not working

- Sync issues

- Quick troubleshooting

It's a reference chapter—read it if you're having problems, or skim it now so you know what's there.

The Big Picture

You've come a long way:

- From scattered, weak passwords to organized, strong ones

- From "I forgot my password" to one-click login

- From texting passwords to secure sharing

- From vulnerability to security

15 minutes per month keeps all of this working perfectly.

That's a tiny investment for massive peace of mind.

Set that calendar reminder right now. First Sunday of every month. 15 minutes. Done.

You've got this!

Chapter 9: When Things Go Wrong (Quick Fixes)

This Chapter Is Your Emergency Guide

Something not working? Don't panic. Most problems are simple to fix.

How to use this chapter:

- Find your problem in the list

- Try the solutions in order

- Usually fixed in 2 minutes

Let's solve some problems.

Login Problems

"I Can't Log Into Bitwarden"

Problem: Wrong password error, can't get in.

Try these in order:

1. Check Caps Lock

- Obvious but common

- Is Caps Lock on?

2. Try Typing Slowly

- Auto-correct might be changing your password

- Type each character carefully

3. Try a Different Device

- Can you log in on your phone vs. computer?

- If yes, problem is device-specific

4. Check Your Master Password Backup

- Where did you write it down?

- Safe? Drawer? Safe deposit box?

5. Try Password Hint

- Login screen → "Get master password hint"

- Check your email

6. Use Emergency Access

- If you set it up

- Contact your trusted person

- They can help you regain access

7. Last Resort

- If you truly forgot your master password

- And have no backup

- You're locked out

- Create new account, start over

Prevention: Write down master password in safe place TODAY.

"My 2FA Code Isn't Working"

Problem: Entering the 6-digit code, but it's rejected.

Try these:

1. Wait for New Code

- Codes change every 30 seconds
- You might be using expired code
- Wait for the next one

2. Check Time on Phone

- Authenticator apps need correct time
- Settings → Date & Time → Set Automatically

3. Type Manually

- Don't copy/paste the code
- Type the 6 digits by hand

4. Use Recovery Code

- Can't get 2FA working?
- Use one of your recovery codes instead
- Login screen → "Use recovery code"

5. Use Backup 2FA Method

- Set up email 2FA as backup?
- Switch to that method

Still stuck? Email support@bitwarden.com

“I’m Locked Out Completely”

Forgot master password AND lost 2FA device AND no recovery codes?

Unfortunately, you’re locked out permanently. This is by design—zero-knowledge encryption means even Bitwarden can’t help.

Your options:

1. Create new Bitwarden account

2. Import from backup (if you have one)

3. Start fresh (if you don’t)

This is why backups matter.

Auto-Fill Problems

“Bitwarden Isn’t Filling My Password”

Problem: Click on password field, nothing happens.

Try these:

1. Manual Approach

- Click Bitwarden icon

- Find your login manually

- Click it to fill

2. Check You Have It Saved

- Open Bitwarden

- Search for the website

- Is it there?

3. Use Keyboard Shortcut

- Press `Ctrl+Shift+L` (or `Cmd+Shift+L`)

- Sometimes works when clicking doesn't

4. Update the Website URL

- Edit the login

- Make sure website URL matches where you are

- Save and try again

5. Refresh the Page

- Reload the website

- Try auto-fill again

6. Restart Browser

- Close browser completely

- Reopen

- Try again

Still not working? Some websites just don't play nice with password managers. Copy/paste manually.

"Bitwarden Filled the Wrong Password"

Problem: Auto-fill put in a password, but it's the wrong one.

Why this happens: You have multiple accounts for this website (Gmail personal, Gmail work, etc.)

Solution:

1. Click Bitwarden icon (don't use keyboard shortcut)

2. You'll see all matching logins listed

3. Click the correct one

4. Should fill properly now

Fix it permanently:

- Give them clear names: "Gmail - Personal," "Gmail – Work"

- Star your primary account (favorites appear first)

"Auto-Fill Not Working on My Phone"

iPhone:

1. Check Settings

- iOS Settings (not Bitwarden)

- Passwords

→ Password Options

- Make sure Bitwarden is checked

2. Disable iCloud Keychain

- Same settings screen

- Uncheck iCloud Keychain

- Only one password manager at a time

3. Restart iPhone

- Turn it off and on

- Test again

Android:

1. Check Autofill Service

- Android Settings → Passwords & Autofill

- Make sure Bitwarden is selected

2. Try Accessibility Mode

- Bitwarden app → Settings

- Auto-fill Services

- Enable "Use Accessibility"

- Grant permission

3. Disable Battery Optimization

- Settings → Apps → Bitwarden

- Battery → Don't Optimize

4. Restart Phone

Sync Problems

"Changes Aren't Showing Up on Other Devices"

Problem: Updated password on computer, but phone still shows old one.

Quick fix:

1. Manual Sync Both Devices

- Device 1: Bitwarden → Settings → Sync Vault Now

- Device 2: Bitwarden → Settings → Sync Vault Now

- Wait 30 seconds

- Check again

2. Check Internet

- Are both devices online?

- Try loading a website

3. Re-Login

- Log out of Bitwarden

- Close app completely -

Log back in

- Should sync fresh

4. Check It's Really Different

- Maybe you changed wrong copy

- Check website to see which password actually works

"Duplicate Passwords Appearing"

Problem: Same login shows up twice (or more).

Solution:

1. Compare Them

- Open each duplicate
- Which one is correct/more recent?
- Which has better info?

2. Delete the Wrong One

- Open the duplicate
- Edit → Delete
- Confirm

3. Sync

- Manually sync
- Check on other devices
- Should be gone everywhere

Mobile App Problems

"App Crashes When I Open It"

Try these:

1. Force Close and Reopen

- Don't just minimize
- Fully close the app
- Reopen

2. Restart Phone

- Turn off

- Turn on

- Try app again

3. Update the App

- App Store or Play Store

- Check for updates

- Install if available

4. Reinstall

- Delete Bitwarden app

- Restart phone

- Reinstall from app store

- Log in (all your passwords are safe on servers)

"Biometric Unlock Stopped Working"

Problem: Face ID / Touch ID / Fingerprint not working.

Try these:

1. Re-enable It

- Bitwarden → Settings → Security

- Disable "Unlock with Face ID" (or Touch ID/Biometrics)

- Close app completely

- Reopen

- Re-enable it

- Try again

2. Test Biometric Elsewhere

- Try unlocking your phone
- Try another app
- If biometric fails everywhere, it's a phone issue

3. Re-register Biometric

- Phone settings
- Face ID / Touch ID / Biometric
- Remove and re-add
- Try Bitwarden again

Browser Extension Problems

"Extension Disappeared"

Problem: Bitwarden icon is gone from toolbar.

Solution:

Chrome/Edge:

1. Click puzzle piece icon
2. Find Bitwarden
3. Click pin icon
4. Should reappear

Firefox:

1. Menu → Add-ons

2. Find Bitwarden

3. Make sure it's enabled

"Extension Is Slow"

Try these:

1. Close Other Tabs

- Too many tabs = slow browser

- Close what you don't need

2. Restart Browser

- Close completely

- Reopen

3. Update Extension

- Browser should update automatically

- Or reinstall from web store

4. Clear Browser Cache

- Browser settings → Privacy → Clear browsing data

- Don't check passwords (just cache)

Getting More Help

Bitwarden Support

Email: support@bitwarden.com

Include:

- What you were trying to do

- What happened instead

- Device and browser (e.g., "iPhone 13, Safari")

- Bitwarden version (Settings → About)

- Screenshots if helpful

Never include:

- Your master password

- Actual passwords from your vault

- Recovery codes

Community Forum

Visit: community.bitwarden.com

- Search for your problem first
- Someone probably had it before
- Post if you don't find an answer
- Community is helpful

This Book's Companion Website

(If you got this as part of a series)

Check the book's website for:

- Updated troubleshooting

- Video guides

- FAQ

- Reader Q&A

The Universal Fix

When in doubt, try this sequence:

1. **Restart the app** (close completely, reopen)
2. **Sync manually** (Settings → Sync)
3. **Restart your device** (phone or computer)
4. **Update Bitwarden** (check for updates)
5. **Log out and back in**
6. **Reinstall** (last resort)

This fixes 90% of problems.

Quick Reference: Most Common Issues

Problem	Quick Fix
Can't log in	Check caps lock, try password backup
2FA code rejected	Wait for new code, check phone time
Password won't fill	Click Bitwarden icon, select manually
Changes not syncing	Manual sync on both devices
App crashes	Restart phone, reinstall if needed
Biometric unlock fails	Disable and re-enable in settings

When Nothing Works

Don't panic.

Worst case:

1. Your passwords are safe (encrypted on Bitwarden servers)

2. You can access web vault (vault.bitwarden.com) from any browser

3. Your data isn't lost

4. Most problems are temporary

Immediate action:

1. Try accessing from web vault

2. If that works, problem is with specific app/device

3. Export a backup while you troubleshoot

4. Contact Bitwarden support

You're Done!

That's it. That's the book.

You now know:

- Why password managers matter

- How to set up Bitwarden

- How to use it daily

- How to stay secure

- How to share with family

- How to fix problems

What to do now:

1. Set up your monthly security check reminder

2. Make sure your master password is written down safely

3. Create your first quarterly backup

4. Keep using Bitwarden every day

5. Help a friend/family member set it up

You're more secure than 95% of people on the internet.

That's worth celebrating.

Appendix: Quick Start Summary

Already know the basics? Here's the express version:

Day 1: Setup (30 minutes)

1. Create account at bitwarden.com

2. Make strong master password (passphrase method)

3. Write it down, store safely

4. Enable 2FA with authenticator app

5. Save recovery codes

Day 2: Install (15 minutes)

1. Install browser extension

2. Install mobile app

3. Enable auto-fill

4. Set up biometric unlock

Day 3: Populate (30 minutes)

1. Import from browser
2. Add important accounts manually
3. Use password generator going forward

Day 4-7: Practice

1. Use Bitwarden for every login
2. Generate passwords for new accounts
3. Build the habit

Monthly: Maintain (15 minutes)

1. Check for weak/reused passwords
2. Check haveibeenpwned.com
3. Change compromised passwords

Quarterly: Backup (10 minutes)

1. Export vault
2. Store encrypted or in safe
3. Delete old backup

That's it. You're done.

Final Thoughts

Password security doesn't have to be complicated.

You don't need to:

- Understand encryption algorithms
- Be a computer expert
- Spend hours managing passwords
- Stress about security

You just need to:

- Use Bitwarden
- Follow the basics in this book
- Spend 15 minutes per month
- Keep using it

The result:

- Stress-free logins
- Strong, unique passwords everywhere
- Protection against breaches
- Family can securely share passwords
- Peace of mind

Welcome to better password security.

You've got this!

Continue the Journey with our Security Books

More Security Books by Prairie Light Press

All titles are part of the **Personal Security in a Digital World** *series.*

Bitwarden Essentials

A practical guide to securing and managing your passwords using Bitwarden, with step-by-step instructions for setup, daily use, and account recovery.

Guard Your Money ***(Recommended Next)***

A clear, practical guide to protecting financial accounts using YubiKey hardware security, covering phishing, identity theft, and modern account takeover threats.

Network Security Essentials

A plain-language guide to securing a home network using Firewalla Gold Plus, focused on understanding security choices rather than just technical settings.

Firewalla: After Setup

Firewalla: After Setup focuses on the decisions that matter once your network is running, how to structure devices, reduce noise, contain risk, and keep security effective over time — without constant tweaking.

Other Books by Prairie Light Press

The Awareness Series

The Awareness Series explores the nature of thought, attention, and the sense of self through direct observation—not theory.

Written in clear, simple language, these books are designed to be **experienced, not analyzed**, helping readers notice awareness as it unfolds in real time.

Awareness (Start Here)

Noticing the Voice Inside Your Head
The starting point — recognizing the voice and stepping out of automatic identification with it.

Still Noticing
The continuation — living with awareness over time, including the parts that are often left out.

These books can be read in any order, but they follow a natural progression:

Awareness → Still Noticing

Explore the Complete Prairie Light Press Catalog

Explore the Complete Catalog

Visit Prairie Light Press
www.prairielightpress.com

Browse on Amazon
Amazon Author Page

Prairie Light Dispatch

Short observations on awareness, daily life, and the systems we build around them.

Quiet notes from the prairie.

Visit:
lgphil7.substack.com

Book Preview

Guard Your Money
The Simple YubiKey Guide for Financial Security

Continue reading with this exclusive preview of **Guard Your Money***: The Simple YubiKey Guide for Financial Security*

Guard Your Money

The Simple YubiKey Guide for Financial Security

PRAIRIE LIGHT PRESS
ESTD. 2025

Helping readers move from the unknown to the known

Contents of the Complete Book
Page numbers refer to the full edition.

Contents

The Simple YubiKey Guide for Financial Security 1

Chapter 1: Understanding the Threats 12

Chapter 2: What Is a YubiKey? (In Plain English) 20

Chapter 3: Getting Started: Your First YubiKey 32

Chapter 4: Setting Up Your YubiKey (Step-by-Step) 42

Chapter 5 — Everyday Life With Your YubiKey 56

Chapter 6 — Backup Keys, Emergency Planning, and Account Recovery 71

Chapter 7 — Protecting Your Entire Family With YubiKeys 86

Chapter 8 — Cleaning Up Old Security Methods (Eliminating Weak Links) 101

Chapter 9 — The YubiKey Troubleshooting & Survival Guide 116

Chapter 10 — The Ultimate Financial Security Upgrade (Beyond YubiKey) 132

Chapter 11 — Advanced YubiKey Tips, Tricks & Expert-Level Secrets 150

Chapter 12 — Your Unhackable Life (Putting It All Together) 165

Appendix A — Quick Start Guides, Checklists & Templates 180

Appendix B — Visual Diagrams & Reference Illustrations 187

Appendix C — Printable Worksheets & Forms 193

Introduction: Why Your Money Isn't Safe (And How to Fix It)

Let me ask you something: How much money do you have in your bank account right now? Your investment accounts? Your retirement savings?

Now imagine waking up tomorrow morning, checking your phone, and seeing that it's all gone. Every penny. Transferred to someone you've never met, in a country you've never visited.

Sound like a nightmare? It happened to Sarah, a 52-year-old teacher from Ohio. She thought she was logging into her bank's website—it looked identical to the real thing. She entered her username and password like she'd done a thousand times before. Within 30 minutes, $43,000 was gone from her savings account.

Or how about Tom, a retired veteran who had his entire retirement account at Schwab drained after someone intercepted a text message code sent to his phone. The thieves used something called 'SIM swapping' to take over his phone number, and the bank's text message security didn't stand a chance.

These aren't rare cases. In 2024 alone, Americans lost over $12.5 billion to online fraud and cybercrime. That's billion with a 'B.' And it's getting worse every single day.

The Problem Is Simple

Your passwords aren't enough anymore. Even if you use a different password for every account (and let's be honest, most people don't), hackers have gotten incredibly sophisticated. They can:

- Send you emails that look EXACTLY like they're from your bank
- Create fake websites that are pixel-perfect copies of the real thing
- Intercept the 'security codes' sent to your phone via text message
- Install invisible malware on your computer that records everything you type
- Hijack your phone number so THEY receive your security codes

The scariest part? You can do everything right—use strong passwords, be careful about emails, keep your software updated—and STILL get hacked. Because the system itself is broken.

But There's a Solution

This book is about one simple device that changes everything: the YubiKey.

Think of a YubiKey like a physical key for your online accounts. It's a small device, about the size of your car key fob, that you plug into your computer or tap on your phone. Without it, nobody—and I mean NOBODY—can access your money, even if they have your password.

- Not the hacker in Russia who phished your password.
- Not the scammer who cloned your phone.
- Not even the sophisticated criminal who installed malware on your laptop.

If they don't have your physical YubiKey in their hands, they can't get to your money. Period.

Who This Book Is For

This book is for regular people who want to protect their money without becoming computer experts. You don't need to understand encryption algorithms or cryptographic protocols. You just need to follow simple, step-by-step instructions.

If you:

- Have money in a bank account, investment account, or retirement fund
- Want to sleep better at night knowing your finances are protected
- Are tired of worrying about getting hacked or scammed
- Don't want to become a tech expert just to stay safe

...then this book is for you.

What You'll Learn

By the end of this book, you'll know exactly how to:

- Buy the right YubiKey for your needs (they're surprisingly affordable)
- Set it up with your bank, investment accounts, email, and password manager
- Create a backup YubiKey and store it safely (this is crucial!)
- Use your YubiKey every day without any hassle
- Protect your entire family's financial accounts
- Recover if something goes wrong

I'll give you specific, step-by-step instructions for the major financial institutions: Chase, Bank of America, Wells Fargo, Vanguard, Fidelity, Schwab, and JPMorgan Chase. No technical jargon. No confusing terms. Just simple, clear directions that anyone can follow.

A Small Investment for Huge Peace of Mind

A YubiKey costs between $25 and $70 depending on the model. That's it. One-time purchase. No monthly fees. No subscriptions.

Compare that to the billions of dollars lost to cybercrime every year, and it's the easiest decision you'll ever make.

So let's get started. Your money is waiting to be protected, and it's easier than you think.

Turn the page, and let's make your financial accounts virtually unhackable.

Chapter 1: Understanding the Threats

Before we talk about the solution, you need to understand what you're up against. And I promise you, it's scarier than you think—but also more preventable than you might imagine.

Phishing: The Wolf in Sheep's Clothing

You've probably heard the term 'phishing' before. But let me show you how sophisticated it's become.

Here's what happened to Jennifer, a marketing director in Seattle:

- She received an email that appeared to be from Chase Bank. The logo was perfect. The formatting was identical. The email address looked legitimate: 'security@chase-alerts.com'. The message said there was suspicious activity on her account and she needed to verify her identity immediately.

- She clicked the link. The website looked EXACTLY like Chase's real website. She entered her username and password. Within seconds, the real criminals had her credentials and were logging into her actual account.

- The kicker? The fake website even showed her real account balance (they grabbed it in real-time after she logged in), so she had no idea anything was wrong. By the time she realized what happened, $18,000 had been wire transferred out of her account.

⚠ WARNING: Modern phishing emails are nearly impossible to detect by eye. Even security professionals get fooled.

Malware: The Silent Thief

Malware is malicious software that gets installed on your computer without you knowing. It can:

- Record every keystroke you type (including passwords)
- Take screenshots of your banking sessions
- Modify websites you visit to steal your information

- Run completely invisibly in the background for months

How does it get on your computer? Sometimes from shady websites, but often from sources you trust:

- Fake software updates that look legitimate
- Email attachments from addresses that appear to be from friends
- Infected USB drives
- Even compromised legitimate websites you visit regularly

SIM Swapping: When Your Phone Number Betrays You

This one is particularly nasty because it exploits a security measure that's supposed to PROTECT you: text message verification codes.

Here's how it works:

- A scammer calls your cell phone carrier pretending to be you
- They claim they 'lost' their phone and need to transfer your number to a new SIM card
- They provide just enough personal information (often gathered from social media or data breaches) to convince the carrier they're really you
- The carrier transfers your phone number to a SIM card that the criminal controls
- Now when your bank sends a 'security code' via text message, the criminal receives it—not you
- They drain your accounts while you're left wondering why your phone suddenly stopped working

This happened to Michael Terpin, a cryptocurrency investor, who lost $24 million in a single SIM swap attack. But it's not just happening to rich people—everyday Americans lose thousands of dollars to this scam every single day.

Password Database Breaches

Even if YOU do everything right with your passwords, the companies you trust might not.

- Major data breaches in recent years:
- Yahoo: 3 billion accounts
- LinkedIn: 700 million accounts
- Adobe: 153 million accounts
- Home Depot: 56 million payment cards

When these breaches happen, criminals get access to usernames, passwords, email addresses, and sometimes even credit card information. Then they try those same credentials on OTHER websites, betting (correctly) that most people reuse passwords.

If you used the same password for LinkedIn that you use for your bank? They're in.

The Shocking Statistics

Let's talk numbers, because they're staggering:

- 81% of data breaches are caused by weak, reused, or stolen passwords (Verizon, 2024)
- Americans lost $12.5 billion to cybercrime in 2024
- The average cost of a data breach for a company: $4.45 million
- 95% of cybersecurity breaches are caused by human error
- A new phishing website is created every 20 seconds

And here's the stat that should really wake you up: The FBI reports that people over 60 lost $3.4 billion to internet scams in 2023 alone. Criminals specifically target older Americans because they tend to have more savings and may be less familiar with online security.

Why Traditional Security Doesn't Work Anymore

'Strong passwords' aren't enough. Even if you create a password like 'Xk#9mP$2qL@7vN!5' and never write it down, it can still be:

- Phished (you type it into a fake website)

- Keystroke-logged by malware
- Stolen in a database breach
- Intercepted if the website doesn't use proper encryption

'Security questions' are a joke. What was your first pet's name? Your mother's maiden name? The street you grew up on? This information is often publicly available on social media or can be easily guessed.

'Text message codes' can be intercepted through SIM swapping, as we discussed. They're better than nothing, but they're not secure enough for your money.

'Authenticator apps' (like Google Authenticator) are better than text messages, but they can still be compromised if your phone is stolen or hacked.

So What DOES Work?

Physical security keys like the YubiKey.

Here's why they're virtually unhackable:

- Phishing doesn't work – Even if you enter your password on a fake website, the scammer can't use it without your physical key

- Malware doesn't work – Recording your keystrokes is useless if they also need the physical device in your pocket

- SIM swapping doesn't work – Your phone number is irrelevant when you're using a hardware key

- Database breaches don't matter – Even if a company's password database is stolen, attackers still can't access your account without your key

Think of it this way: A burglar might pick your door lock, but they can't pick a lock that doesn't exist. With a YubiKey, there IS no lock for them to pick remotely. They would need to physically steal the device from you—and even then, they'd still need your password.

💡 TIP: Security experts and major corporations use hardware keys for a reason. Google has been using them internally since 2017 and reports ZERO successful phishing attacks against employees who use them.

In the next chapter, we'll talk about exactly what a YubiKey is and how it works. And don't worry—it's way simpler than you think.

Chapter 2: What Is a YubiKey? (In Plain English)

Alright, let's demystify this thing. If you've never heard of a YubiKey before, or if you have but you're not quite sure what it does, this chapter will make everything crystal clear.

The Car Key Analogy

- Think about your car key. Even if someone knows:
- Your car's make and model
- Where you park it
- Your license plate number
- That you always keep the owner's manual in the glove box

...they STILL can't drive your car without the physical key, right?

- A YubiKey works the same way for your online accounts. Even if hackers know:

- Your username
- Your password
- Your email address
- Your phone number

...they STILL can't access your account without the physical YubiKey that's in your pocket (or wherever you keep it safe).

What Does It Look Like?

Figure 10 - YubiKey on Keychain -LP/PLP

A YubiKey is a small device, roughly the size of your car key fob or a USB flash drive. Most models are about 1.5 inches long and less than an inch wide. It's durable (you can literally put it through the washing machine and it'll still work), waterproof, and has no battery to charge.

The most common model has:

- A USB connector (either USB-A or USB-C, just like a regular USB stick)
- A gold circle or square on the flat surface (this is what you touch to activate it)
- Optional NFC capability (for tapping on your phone)

There are no buttons to press, no screens to read, and nothing complicated about it. It's literally just a small piece of hardware that you plug in or tap when asked.

How It's Different From Text Message Codes

You might be thinking, 'But my bank already sends me security codes via text message. Isn't that good enough?'

Short answer: No. Here's why:

When your bank sends you a code via text message, that code is:

- Traveling through the air (can be intercepted)
- Stored on your phone (can be accessed if your phone is hacked)
- Tied to your phone number (can be redirected via SIM swapping)
- Vulnerable to social engineering attacks on your carrier

With a YubiKey:

- Nothing travels through the air—the authentication happens directly between the key and the website
- Nothing is stored anywhere—each authentication generates a unique, one-time cryptographic signature
- Your phone number is irrelevant—the key doesn't use it
- Nobody can call your carrier and trick them—there's no carrier involved

TIP: The National Institute of Standards and Technology (NIST) recommends deprecating SMS-based authentication due to security vulnerabilities. Hardware keys like YubiKey are their recommended alternative.

The Science Made Simple (No Geek Speak, I Promise)

Okay, I know you didn't buy this book for a computer science lesson, but understanding the basics helps you trust the system. I'll keep this REALLY simple.

When you set up a YubiKey with, say, your bank:

- The YubiKey and the bank exchange mathematical 'handshakes' (called cryptographic keys)
- These handshakes are unique to YOUR YubiKey and YOUR bank account
- Every time you log in, your YubiKey proves it's the same device without revealing any secret information
- Even if hackers intercept the communication, they can't copy or reuse it

It's like having a secret handshake with your bank that changes every single time and can't be faked. Nobody else knows the handshake, and even if they watch you do it, they can't recreate it.

The beautiful part? You don't have to understand how it works. You just plug it in (or tap it), touch the gold circle, and you're authenticated. That's it.

Why Hackers Can't Fake It

Here's the crucial part: Unlike a password (which is just information that can be copied), a YubiKey uses physical hardware that generates unique cryptographic signatures.

To fake a YubiKey, a hacker would need to:

- Steal or copy the mathematical secrets stored in the hardware (essentially impossible without physical access and specialized equipment)
- Break the cryptographic algorithms (would take longer than the age of the universe with current computers)
- Somehow predict the unique signature it will generate next (mathematically impossible)

Even nation-state actors with unlimited resources can't break this. That's why the U.S. government, major tech companies, and high-security organizations all use hardware security keys.

Which YubiKey Should You Buy?

Yubico (the company that makes YubiKeys) offers several models. Don't let the choices overwhelm you—I'll break it down simply.

The YubiKey 5 NFC – Best for Most People ($45)

This is the model I recommend for almost everyone. Here's why:

- Works with computers (USB-A port, found on most computers)
- Works with phones (NFC means you can tap it on your iPhone or Android)
- Compatible with virtually every service that supports security keys
- Durable and waterproof
- No battery needed—it powers itself from the USB port

Figure 11 - YubiKey 5C NFC - LG/PL

The YubiKey 5C NFC – For Newer Laptops ($55)

Watch for more Prairie Light Press Reader Previews

www.ingramcontent.com/pod-product-compliance
Lightning Source LLC
LaVergne TN
LVHW090354240926
848801LV00002B/34
* 9 7 9 8 9 9 6 9 8 0 2 1 5 *